Statistical Design - ANOVA of Light Emitting Diode Horticulture Fixtures using IBM-SPSS statistical software 2016-2017, Final Analysis

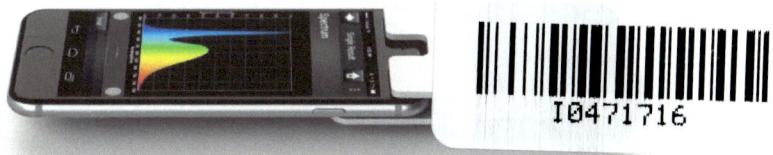

1. Results and Recommendations

2. Final Analysis

3. Daystar Life Center

4. Special Thanks

5. Work Cited

Thank you very much <u>Allied Scientific Pro</u> for you innovative technology and easy to use app! Your Lighting Passport made a complicated project run smooth due to pin point precision.

https://alliedscientificpro.com/

http://www.lightingpassport.com/

There are five major areas of this research. Statistics specifically for business and economics. Physics and the essential understanding of light and how its absorbed by humans and plants. Botany the science of plants. Chemistry and the different aspects of soil culture. And of course electrical engineering which consider of the different sections of a horticulture LED.

There is most definitely a variation between LED fixtures most simply described as white light LED grow lights and Pink light which is comprised of a collection on mostly blue and red LEDs. Other things to consider are whether or not eye protection is required? Do companies market their products honestly? Does the fixture have a visual mode for the human eye so we can examine the plants for nutrient deficiencies, molds, pests, etc.?For this LED ANOVA the accepted hypothesis is as follows;

H1- All participating LED grow lights will not have the same effect on plants

$\mu 1 \neq \mu 2 \neq \mu 3 \neq \mu 4 \neq \mu 5 \neq \mu 6 \neq \mu 7 \neq \mu 8 \neq \mu 9 \neq \mu 10$

For this first phase of the ANOVA test The California Light Works; Solar system 550 was the most economical LED fixture.

www.californialightworks.com

This LED ANOVA Variation started with separating everything into two main categories of LED fixtures which are pink light LEDs and white light LEDs.

Pink	*White*
Ion 8 Lighthouse Hydro	*Solar Pro 900AMARE*
G8-900 DormGrow	*Universe OGlights-3500K*
V2 VIVID GRO	*Solar Eclipse 450 uvb AMARE*
V1 VIVID GRO	*COB Grow Lights – Scorpion9*
Solar System 550- California Light Works	
ALPHA HydrogrowLED	

Pink light population & white light population

Lighting Passport Results McCree

Solar System 550- California Light Works

Center Measurement   Corner Measurement

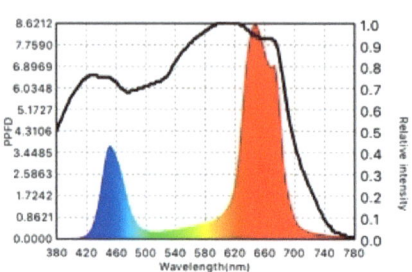
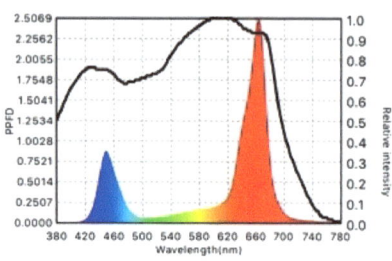

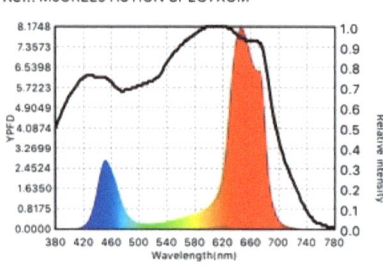
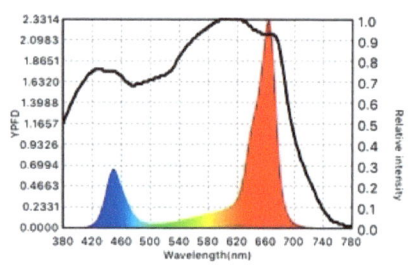

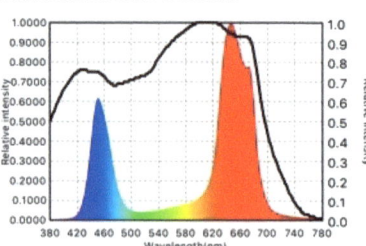
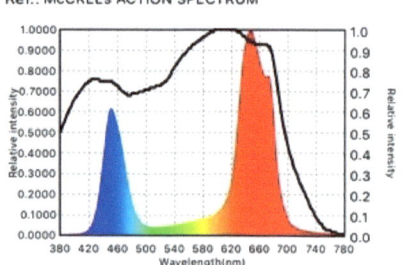

CIE1931

x:0.3639, y:0.1822   CCT: 1814 K

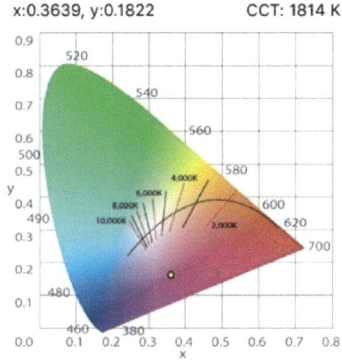

CIE1931

x:0.3411, y:0.1784   CCT: 2221 K

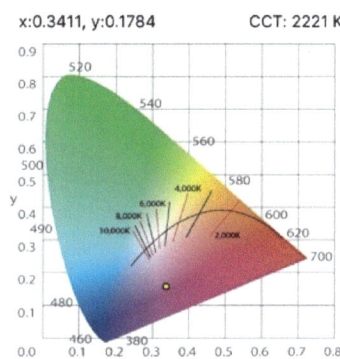

CIE1976

u':0.3265, v':0.3678   CCT: 1814 K

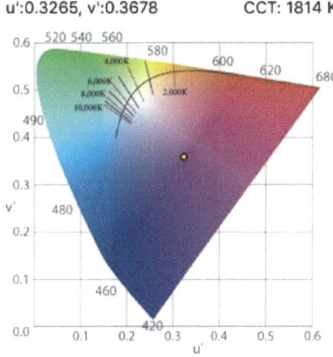

CIE1976

u':0.3061, v':0.3601   CCT: 2221 K

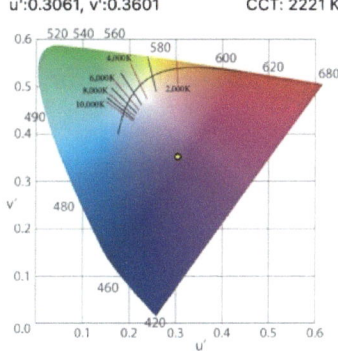

Parameter	Value	Parameter	Value
PPFD (400~700 nm)	671.16 μmol/m²s	PPFD (400~700 nm)	140.34 μmol/m²s
PPFD IR (701~780 nm)	18.790 μmol/m²s	PPFD IR (701~780 nm)	3.1228 μmol/m²s
PPFD R (600~700 nm)	483.29 μmol/m²s	PPFD R (600~700 nm)	98.375 μmol/m²s
PPFD G (500~599 nm)	46.448 μmol/m²s	PPFD G (500~599 nm)	11.611 μmol/m²s
PPFD B (400~499 nm)	141.51 μmol/m²s	PPFD B (400~499 nm)	30.363 μmol/m²s
PPFD UV (380~399 nm)	0.5851 μmol/m²s	PPFD UV (380~399 nm)	0.1291 μmol/m²s
YPFD (400~700 nm)	589.86 μmol/m²s	YPFD (400~700 nm)	123.50 μmol/m²s
YPFD (380~780 nm)	594.30 μmol/m²s	YPFD (380~780 nm)	124.25 μmol/m²s
YPFD IR (701~780 nm)	4.1061 μmol/m²s	YPFD IR (701~780 nm)	0.6744 μmol/m²s
YPFD R (600~700 nm)	445.54 μmol/m²s	YPFD R (600~700 nm)	90.935 μmol/m²s
YPFD G (500~599 nm)	41.352 μmol/m²s	YPFD G (500~599 nm)	10.349 μmol/m²s
YPFD B (400~499 nm)	103.03 μmol/m²s	YPFD B (400~499 nm)	22.231 μmol/m²s
YPFD UV (380~399 nm)	0.3455 μmol/m²s	YPFD UV (380~399 nm)	0.0761 μmol/m²s
R/ B	3.42	R/ B	3.24
R/ FR	25.72	R/ FR	31.50
DLI	57.989 mol/m²	DLI	12.125 mol/m²
Illuminance	15356 lux	Illuminance	3135 lux
λp (380~780 nm)	647 nm	λp (380~780 nm)	663 nm
λD (380~780 nm)	0 nm	λD (380~780 nm)	0 nm
CCT	1814 K	CCT	2221 K
CRI	-6	CRI	6

Lighting Passport Results in Medical Curve

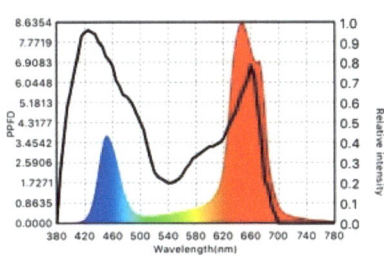

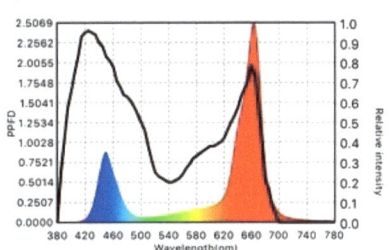

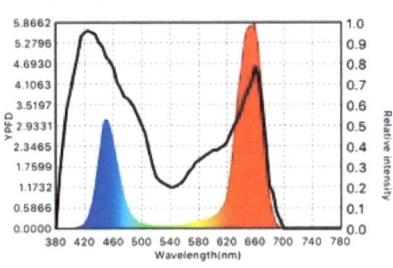

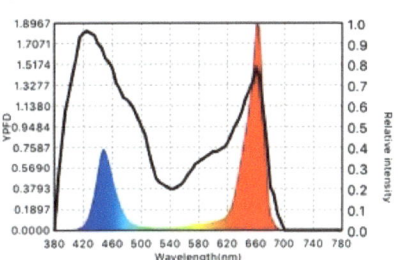

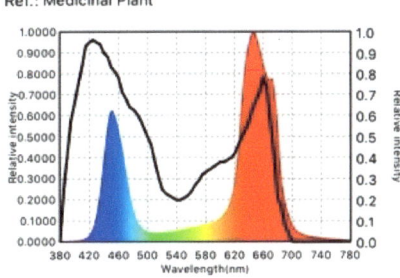

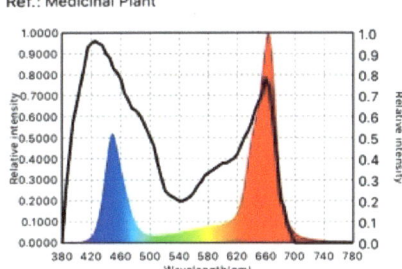

CIE1931

x:0.3655, y:0.1822   CCT: 1792 K

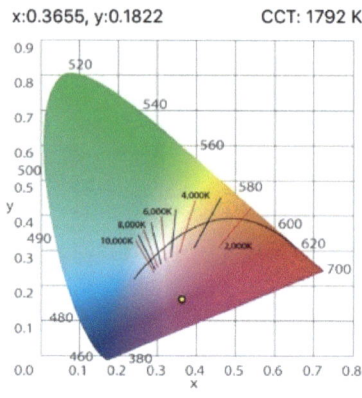

CIE1931

x:0.3411, y:0.1784   CCT: 2221 K

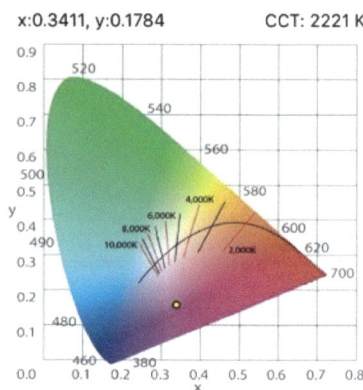

CIE1976

u':0.3281, v':0.3681   CCT: 1792 K

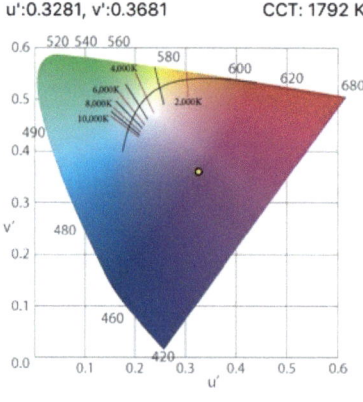

CIE1976

u':0.3061, v':0.3601   CCT: 2221 K

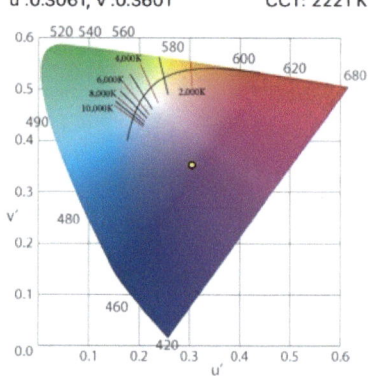

Parameter	Value	Parameter	Value
PPFD (400~700 nm)	672.14 μmol/m²s	PPFD (400~700 nm)	140.34 μmol/m²s
PPFD IR (701~780 nm)	17.934 μmol/m²s	PPFD IR (701~780 nm)	3.1228 μmol/m²s
PPFD R (600~700 nm)	483.24 μmol/m²s	PPFD R (600~700 nm)	98.375 μmol/m²s
PPFD G (500~599 nm)	46.632 μmol/m²s	PPFD G (500~599 nm)	11.611 μmol/m²s
PPFD B (400~499 nm)	142.34 μmol/m²s	PPFD B (400~499 nm)	30.363 μmol/m²s
PPFD UV (380~399 nm)	0.5808 μmol/m²s	PPFD UV (380~399 nm)	0.1291 μmol/m²s
YPFD (400~700 nm)	387.80 μmol/m²s	YPFD (400~700 nm)	84.742 μmol/m²s
YPFD (380~780 nm)	388.04 μmol/m²s	YPFD (380~780 nm)	84.791 μmol/m²s
YPFD IR (701~780 nm)	0.0000 μmol/m²s	YPFD IR (701~780 nm)	0.0000 μmol/m²s
YPFD R (600~700 nm)	261.59 μmol/m²s	YPFD R (600~700 nm)	56.903 μmol/m²s
YPFD G (500~599 nm)	13.965 μmol/m²s	YPFD G (500~599 nm)	3.4406 μmol/m²s
YPFD B (400~499 nm)	112.21 μmol/m²s	YPFD B (400~499 nm)	24.388 μmol/m²s
YPFD UV (380~399 nm)	0.2362 μmol/m²s	YPFD UV (380~399 nm)	0.0535 μmol/m²s
R/ B	3.39	R/ B	3.24
R/ FR	26.95	R/ FR	31.50
DLI	58.073 mol/m²	DLI	12.125 mol/m²
Illuminance	15573 lux	Illuminance	3135 lux
λp (380~780 nm)	647 nm	λp (380~780 nm)	663 nm
λD (380~780 nm)	0 nm	λD (380~780 nm)	0 nm
CCT	1792 K	CCT	2221 K
CRI	-5	CRI	6

The goal of this project was to run a ANOVA that would identify the most economical LED fixture. The next goal was to find out which could most efficiently replacement for a **1000w HID** system MH & HPS. A more economical replacement was found for a slightly higher price. When all ANOVA variables are considered especially debates, cooling costs, and life hours; it's pretty straight forward. Right before starting to cultivate the tomato plants we took a look at all the statistics collected with the watt meter, spectrometer, and agriculture app. This Excel spread sheet is a great starting point for a serious investor building facilities.

In early 2017 the US Senate passed **Bill 518** of the energy conservation standards exemptions safe and efficiently remove obstacles to innovation, research and development, and addresses unintended consequences from the previous bills on energy, specifically LED lighting. LEDs on average are 80% more efficient than fluorescents and these are figures from the national energy department. Continental improvements are forecasted in the LED industry for example Osram will release a new line of emitters for LED horticulture fixtures. Although it may take a additional six months before we actually see these fixtures with the new 2017 Osram emitters hit the market we can see how continuous innovation is sewn into the fabric of the industry.

Solanaceae-Solanaceous

Tomatoe-Solanum Lycopersicum

Tomatoes can be a difficult crop to cultivate as there is no nectar in the flowers, and therefore totally unattractive to honey bees. This ANOVA helped us realize that there is no one size fits all for LED horticulture applications. Since there was a significant amount of data collected we can still use SPSS to run the overall regression and see how things came out.

At first I thought this ANOVA which was designed to help investors seek clarity actually began to foggy the water, up until a certain point where everything clicked right into place. The lighting passport spectrometer and easy to use software really made this project possible, and that's why I've allowed Allied Scientific Pro to publish the research. After studying LED technology for over a year at a 4000 level I feel confident that I can accurately answer any questions that may come up if there is something worth discussing if further detail. The previous paper explains much of the pretext to this final analysis.

<u>Indeterminate Trellising System</u>

This ANOVA is not only internet research, but a collection of academic papers, scholar.google.com investigation hours along with countless hours inside the university library. After the lab portion and implementation of the design the research is crystal clear in the statistical findings and the lab makes things even clearer as we can observe the floral developments fom each individual LED treatment. After running the ANOVA regression in SPSS its recommended that anyone leveraging LED technology into a facility should go with the Solar System 550 from California Light Works.

After consulting with multiple companies like our participants and outsind companies like Lumigrow for example; We came to find several aspects of a LED grow to be true. One is the fact that efficiency of the entire process can be increased by creating a environment in which the plants can sence light from multiple different areas. These LEDs create synergy amongst eachother allowing the corner areas of weak intensity to be converted to optimum areas and the plants thrive in a greater area.

White light LEDs operate in what seems to be the market place where HID growers are searching for ways to improve, but don't want to take the leap of faith into the pink light LED area. This is completely understandable with the amount of misinformation on the internet in general, and when we also consider the history of the LED horticulture sector and the promises one can easily understand why professional agriculturists are hesitant to switch to this new technology.

Rooting Characteristics of Indeterminate F1

The white light LEDs seem to operate in the area of the market where one may not have the knowledge and know how to successfully harvest a crop from pink light LEDs, and clearly not everyone can stop what they're doing to explore the never-ending realm of LED knowledge. So the White light LEDs are like a bridge connecting the two sectors together HID-White LED-Pink LED. Pink LEDs optimize LED technology in general by focusing all energy to the two primary areas of photosynthesis which are 440nm and 660nm. The red 660nm portion is a higher amount because the Blue 440nm is better for photosynthesis. A higher amount of Red 660nm is requires to match the efficiency of the Blue 440nm section of the spectrum. Certain crops like many tomato varieties are difficult to grow under Pink LED lighting, much more difficult than the HID systems were use too and yet still more difficult that white light LEDs. For example here we can see a series of blisters from the intensity of PPFD when the plants were only three weeks old

White light LEDs obviously have a more balanced spectrum and we did not see this effect in those populations. The veg process was smooth and efficient for all White light LED populations. It was the onsite of floral formation and fruit development capability where they came up short.

<u>Below we can see that the sun still has significant values, Although in northern climates LED lighting will be required to keep the price of food moving lower.</u>

Lighting Passport Results in Medical Curve from Sun

PPFD Spectrum

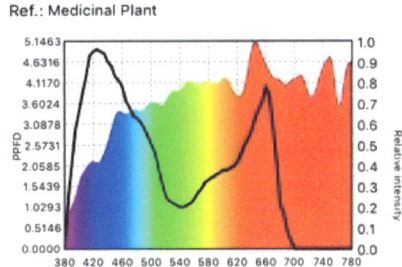

Weighted Spectrum

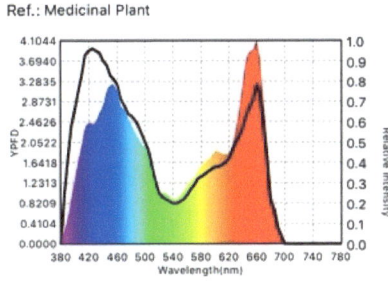
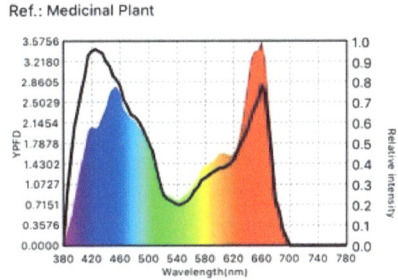

Original Spectrum

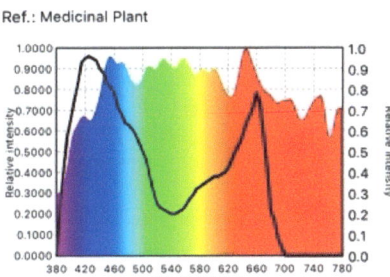
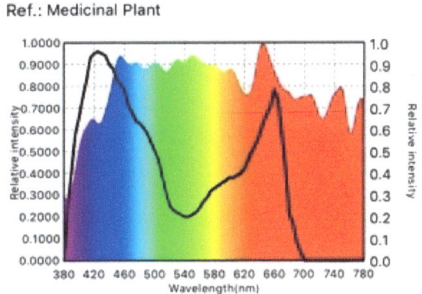

CIE1931

x:0.3340, y:0.3445   CCT: 5434 K

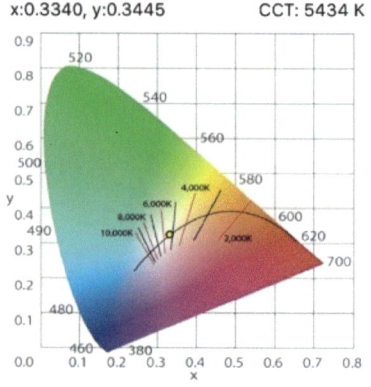

CIE1931

x:0.3343, y:0.3460   CCT: 5423 K

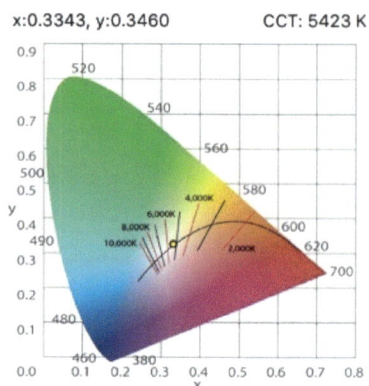

CIE1976

u':0.2066, v':0.4795   CCT: 5434 K

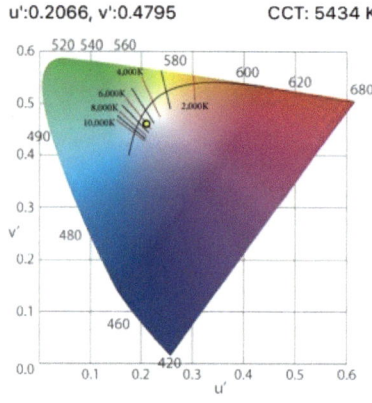

CIE1976

u':0.2063, v':0.4803   CCT: 5423 K

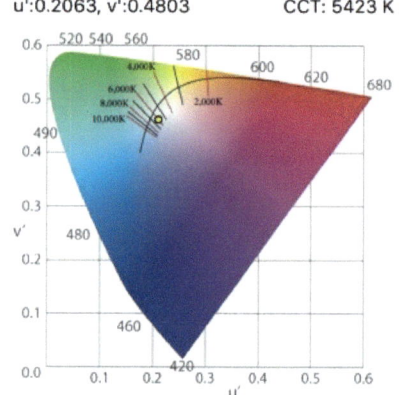

Parameter	Value	Parameter	Value
PPFD (400~700 nm)	1275.0 μmol/m²s	PPFD (400~700 nm)	1113.3 μmol/m²s
PPFD IR (701~780 nm)	375.80 μmol/m²s	PPFD IR (701~780 nm)	336.66 μmol/m²s
PPFD R (600~700 nm)	493.71 μmol/m²s	PPFD R (600~700 nm)	432.05 μmol/m²s
PPFD G (500~599 nm)	453.49 μmol/m²s	PPFD G (500~599 nm)	394.56 μmol/m²s
PPFD B (400~499 nm)	327.76 μmol/m²s	PPFD B (400~499 nm)	286.64 μmol/m²s
PPFD UV (380~399 nm)	26.696 μmol/m²s	PPFD UV (380~399 nm)	22.501 μmol/m²s
YPFD (400~700 nm)	605.07 μmol/m²s	YPFD (400~700 nm)	527.19 μmol/m²s
YPFD (380~780 nm)	616.67 μmol/m²s	YPFD (380~780 nm)	536.98 μmol/m²s
YPFD IR (701~780 nm)	0.0000 μmol/m²s	YPFD IR (701~780 nm)	0.0000 μmol/m²s
YPFD R (600~700 nm)	224.70 μmol/m²s	YPFD R (600~700 nm)	196.37 μmol/m²s
YPFD G (500~599 nm)	132.07 μmol/m²s	YPFD G (500~599 nm)	114.95 μmol/m²s
YPFD B (400~499 nm)	248.31 μmol/m²s	YPFD B (400~499 nm)	215.91 μmol/m²s
YPFD UV (380~399 nm)	11.427 μmol/m²s	YPFD UV (380~399 nm)	9.6424 μmol/m²s
R/ B	1.51	R/ B	1.51
R/ FR	1.31	R/ FR	1.28
DLI	110.16 mol/m²	DLI	96.187 mol/m²
Illuminance	71062 lux	Illuminance	62060 lux
λp (380~780 nm)	645 nm	λp (380~780 nm)	645 nm
λD (380~780 nm)	557 nm	λD (380~780 nm)	558 nm
CCT	5434 K	CCT	5423 K
CRI	99	CRI	99

Results for California Light Works Solar System 550

PPFD (400-700nm) average/w draw   413.552/405w

PPFD (400-700nm) average/$ 413.552/$795

DLI (mol/m^2) average/w draw   35.751/405w

DLI (mol/m^2) average/$ 35.751/$795

YPFD (380-780nm) average/w draw   367.4/405w

YPFD (380-780nm) average/$ 367.3/$795

Average of Efficiency Percentage for McCree Curve   25.39%

Average of efficiency percentage for Medical Curve   34.39%

Average of Red Blue Ratio R/B   3.704%

Spectrum adjustment   yes

LED connectivity option   yes

Remote control ability via smart phone app   no

Fixture Lifespan in hours   50,000 hrs.

Average Heat Output   3.8degrees Fahrenheit

Warrantee 5 years

LED Fixture meets VC funding requirements ANSI/NEC UL certified yes

Weight 13 LBS.

Price/Lifespan $795/50,000

Humidity resistance for Greenhouse application   yes

Rebates from US power companies   yes

California Light Works Solar System 550

The California Light Works solar system 550 and the Dormgrow G8-900 were very similar in terms of spectrum it's appropriate to mention that with the G8-900 there Blue or 460nm measurement was greater than the Solar System 550, and on the other side of this comparison the Red 660nm measurement was greater for the Solar System 550. A noticeable difference that makes working under these lights easy is the view mode on the California Light Works fixture. Not only is there a view mode, but there are two quiet cooling fans and a easy to use light controller that allows for spectrum adjustment, and intensive control. Optimum par output for a 400w fixture made this unit easy to use and adjust.

This controller is a key aspect because when horticulturist are operating multiple units its taxing to be raising and lowering the lights throughout the plants growth stages. A benefit from LED lighting is that they ca n be placed close to the canopy without burning the leafs or creating light bleaching. Also moving forward something that's highly anticipated in the industry is a Smartphone app to control multiple LEDs. The California Light Works controller is certainly a step in the right direction. It also allows for multiple fixtures to be connected together via Ethernet cable which again lowers labor costs and makes spectrum and intensity adjustment simple. These fixtures are UL certified and capable of getting rebates from certain power companies mostly in Washington and California. Since the proper certification was obtained by California Light Works, one can also expect a significantly less insurance policy compared to a HID facility simply because the risk of fire is significantly reduced, yet another hidden advantage of making the change. The Solar System 550 is perfectly capable of producing crops greater than or equal to a 600 HID.

There will be more HID ANOVA tests coming late 2017 and all the HID data can be ran together with the LED data collected which will present a even larger overall perspective of the situation.

Tomato Development from a Red/ Blue treatment

As more and more agriculturist professionals begin to see the benefits of using LEDs the industry is forecasted to grow exponentially up to 2022. All agriculturists should have a firm grasp on their goals, limitations, and requirements before making the switch. Depending on the crop and area, one may find they require something different. With that in mind the industry is very innovative and completive as well, therefore one can expect to see LED improvements even before the "Break Even" point is met. Although LED lighting can improve overall cost of operation through electrical savings, cooling costs, crop improvement, easy of labor ect, its best to continually leverage portions of a large facility. This method works best as adjustments are required and generally speaking its different growing with LED lighting. Also as new improvements come out one can reap the benefit of having cutting edge technology at all times.

For the **F1 Trust** tomatoes provided by Johnny's seeds the Red 660nm was the trick to getting full development. This is why the final analysis emphasizes this so much. Although all statistical ratios and factors should be considered one may also want to ask the company about their 660nm intensity. Without the Lighting Passport spectrometer this would have been very hard to identify, another point that requires mention is that this research suggests plants create a synergetic effect from light; meaning that UV, IR have effects on a plants metabolism, rate of photosynthesis and therefore biomass.

With this in mind, the PhDs referenced in this paper agree that the sun is our most reliable source of energy and should be used whenever possible. Simplicity is the key really and although this research is complicated it can be best for some to take a step back and really thing about what is a grow light and why do we use them when we have a sun? Naturally it's easy to say this coming from Florida, the answer is that not all nations around the world have accurate sunlight to produce a affordable crops. This is where the importance of LED lighting comes into play. As our environments continue to deteriorate and The population continues growing we all should ask ourselves; What can I do to help?

Plants use less green light due to <u>evolutionary in nature</u>. Early on when plants were first starting, photosynthetic Achaea tended to absorb most of the sun's energy in the green spectrum. The result was that plants developed pigments to absorb everything else that wasn't being soaked up by the photosynthetic Achaea. The pigment in Achaea is called <u>bacteriorhodopsin</u>.

The interesting thing about bacteriorhodopsin is that it's a light-activated proton pump. It moves protons across a membrane in response to light. This movement is then converted into chemical energy. This is an inefficient process compared to ATP; because of two energy conversions -- first light energy into mechanical energy, then mechanical energy into chemical energy. Every time you convert energy, you lose a significant amount as heat and generate undesired byproducts.

Tomato Comparison from Three Different Treatments

Chlorophyll, on the other hand, almost directly converts light energy into chemical energy, through ATP which is an efficient process. This is why plants don't need nearly as big a portion of the spectrum, because their process is more efficient. Allied Scientific Pro has just released a new patented Spectral X agriculture net that blocks the majority of green light from entering the greenhouse. There is clearly a rapid rate of innovation in this sector.

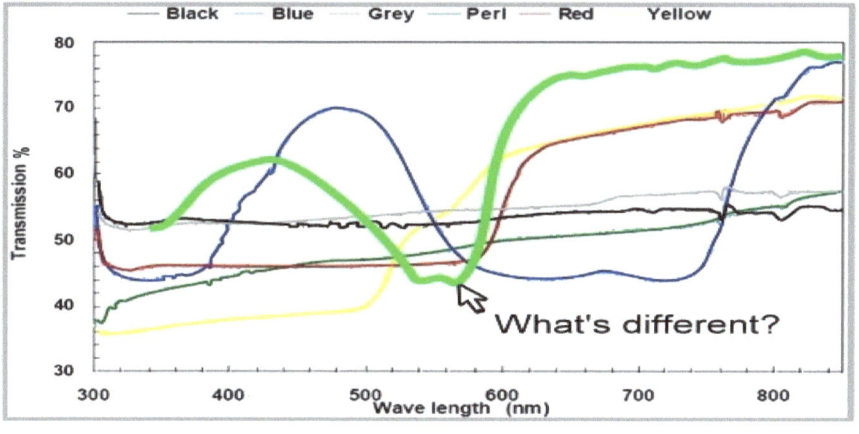

https://alliedscientificpro.com/shop/product/spectral-x-agricultural-net

Green light is still a controversial topic and both methods with and without green have been successful. However we use LEDs to optimize plant production and we must ask ourselves; should we use resources to give our plants green light. We have found that one should not burn resources to give sole lighting projects green light..

This ANOVA suggests that although plants can create synergetic effects from different portions of light, including green, however it's not wise to burn resources to implement the green portion of the McCree curve. The synergetic effects are also influenced by UV and IR light. This is why we now have a Medical curve that shows further research on this topic. One question that came to mind very often was; Why would a LED company have the green portion of light? I asked myself if the PhDs in lighting say that plants don't use green light then why is green light in the white light Spectrum? Although many have suggested a fuller spectrum and they clearly have a place in the market, from a Global Business Management perspective the answer is very simple; Profit.

California Light Works Solar System 550

There are two primary areas of photosynthetic light blue and red, other portions of light influence things like terpines from UV and Phytochrome from IR. Green light can also help the synergetic process of a plant, but the idea behind LED technology is to produce FOOD crops is inexpensive as possible. When we look at food producing companies we see use of sun and pink light LEDs. When we consider moving toward a PFAL for our food in the future, reducing labor costs and optimizing output is essential to lower the cost of produce while simultaneously increasing the nutritional value and benefits one receives from eating these crops. Thus when we consider all data collected from the lighting passport, and the ANOVA lab, and the P3 Watt meter its clear which fixture is the most economical overall.

When we measure light with a spectrometer the measurements in comprise the green portion of light, this is why some argue that PAR is not sufficient method of measuring and the spectrum is actually the key. Yes this is correct and Spectrum is one of the most important aspects of a LED fixture. Investors will see a great return on their time if they call in and ask a few questions about the LED lights they're considering. As these specific questions are asked one can judge from tone of voice how disturbed some get when a spectrum and par measurement are asked for. Unfortunately the LED industry has a history of promising things that could not be delivered, so naturally there is a lot of skepticism coming from both sides. This ANOVA will help clear the water.

Similar to the controversy over green light there is also controversy as to whether or not its ideal to **prune** indeterminate hybrid tomato plants to one vine. After careful consideration of the space we had and the other restrictions we decided to prune the suckers after a strung vegetative phase for eighth weeks. Any small week branch was removed and all new suckers were pruned off while we left any main branch that had already developed. Primary vine and secondary vines produce well but no thirishary vines were allowed to develop. With the drop and lean method frequently used in greenhouse tomato production one would remove all suckers and thus force the plant to grow as one vine continually harvesting the lower clusters and dropping the vine down from the roof while letting the older portion lay on the ground. Unfortunately this was not possible in out 4x4 hydrohut. Since this research was targeted at indoor facilities rather than greenhouses the secondary vine development fit the model much better.

Root Development on Indeterminate Variety

Although I wanted to allow most successful branches to develop the below pictures illustrate exactly why it was necessary to have some kind of controlled production system like drop and lean to optimize the production of the fruit. Although slightly different than drop and lean very similar methods were successfully used in multiple greenhouses here and we decided that was the best fit for this lab.

At the fifth week the plants began a explosive vegetative phase under every light treatment, and it was at this point that spinning or rotating the plants was no longer feasible. A interesting note is that when the transplant was made from two gallon root pouches to five gallon root pouches there was absolutely no visual transplant shock. There was a general level of sensitivity to the LED during the first four weeks, yet afterword the growth rate was fantastic. The notes indicate a fear that the lack of 660nm intensity coming from white light treatments would not be sufficient for these F1 Trust plants to fully develop.

Visual Mode

The white light treatments created great vegetative area, but simply lacked the key element which we found to be **660nm wavelength**, the 660 is the primary area of photosynthesis and this should be a focus for any investors seeking to build a facility powered by LED lighting. Another point of observation is; even though the LED lights were two feet from the canopy there corner measurements are still significantly lower than the center (except for the V1 & V2 from Lighting Science) So, it's a recommendation that if a light mover is a option the grower may want to consider that. See PAR values from ANOVA R/R. The fixtures from Lighting Science have a different design which covers the canopy much better.

Continued Floral Development

Leaf Development from a RGB Treatment

Floral development under a Solar System 550

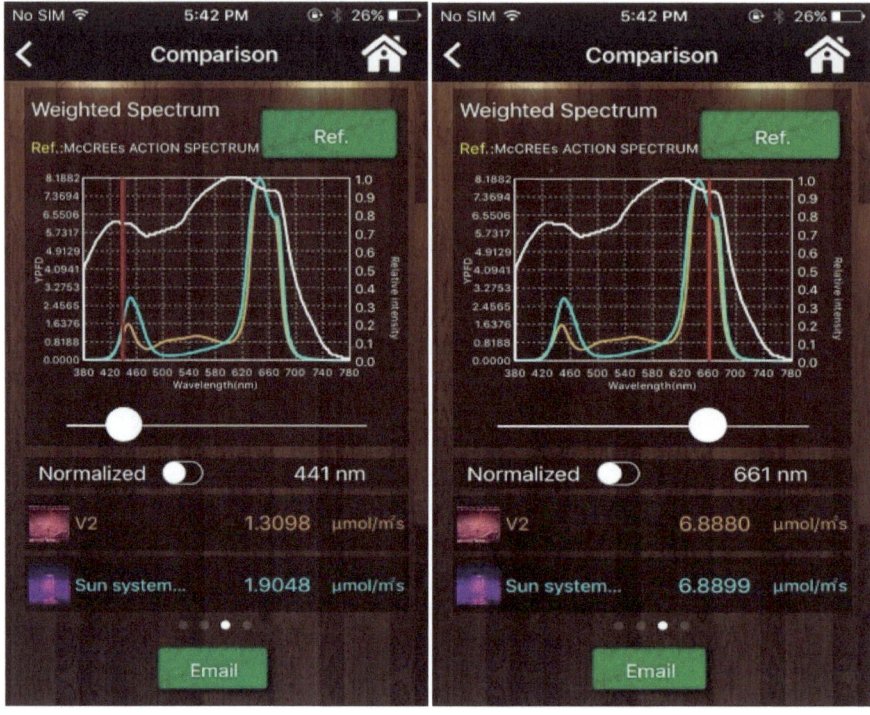

Here in this comparison we can observe overall similar values at the 440nm and 660nm wavelengths. Although one point is that the V2 is a 600watt draw from the wall while the Solar System 550 is only draws400watts. Below we will observe a corner measurement between the V2 and the Solar System 550 and we will see that the coverage capability of the V2 is better that the Solar System 550, however with all variables considered the California Light Works fixture is still a more economical choice.

| Blue 440nm | Red 660nm |

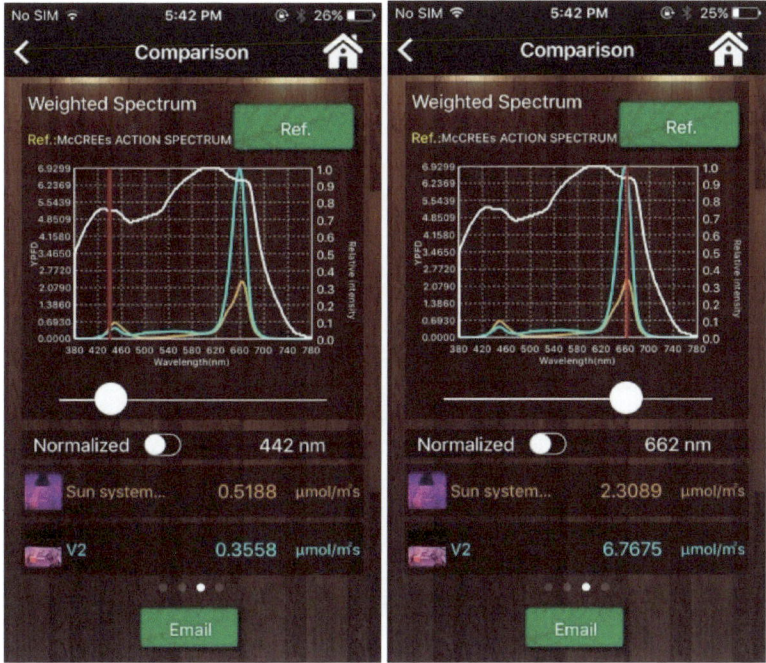

As review the corner measurements from these two fixtures we begin to realize why it's so important to have a firm grasp on the goals and objectives of the project before actually investing

Continued Floral Development

There is a tight equilibrium and balance point over the canopy footprint. Emitters, degree, and angle of the lenses can all change from company to company and this variation aspect should also be matched with the projects goals. The point of leveraging LED technology is to use artificial light to stimulate, influence, and manipulate different aspects of a plants growth rate through its metabolism, therefore producing a more plentiful and nutritious harvest.

During the eighth week of growth the variation among the populations began to show. The flower clusters under the pink light LEDs especially the Solar system 550 and the G8-900 began to rapidly develop. And I mean every day these flower clusters are maturing and developing new clusters. It's important to mention that indeterminate tomato varieties continue fruit development and a form of vegetative growth simultaneously. As the fruit continued developing under the Pink light the white light as a very different story.

The flowers dropped off from either not having enough strength to pollinate, or not having enough light intensify at the 660nm level. Although the science of testing spectra on plants is plant specific this analysis does hit the key elements investors search for. Notice the yellowing of the flower and the lack of vigor, just days before it fell off.

Floral Development Failure

The white light populations continued a explosive vegetative growth phase for two more weeks as over 75% of the flowers fell off. Then at week ten the white light populations continued with intermodal elongation that was very disappointing. The plants begin to lose vigor and started bending over from the internodes being so long in corner and side areas while a light burn set into the center of the population. The photos here clearly illustrate the effects. This research shows that the 660nm is the requirement for fully developing a this variety of tomato.

Here we can again see that there is no one size fits all and investors must take things serious and research what they need to meet their specific goals. Considering goal and objectives is key to a successful implementation of a business model in the long run. This is why the California Light Works Solar System 550 is statistically moving towards being recommended by SPSS software by IBM. And since a PAR/w measurement can't say much about plant health and vigor, yields, and nutrition; The spectrum can, and that is why there is a complete report of the participants spectrums along with all light measurements for the center and corner area at four feet. Also the spectrum accuracy and efficacy.

All populations were given the same treatment except for the lighting. After the first two months of the veg cycle were completed the lighting cycle changed from 16/8 to a 12/12. Indeterminate tomato varieties are ideal for test crops because they cannot be influenced by photoperiod. So since we can't force these plants to flower by lowering the amount of daylight hours, we must allow them to continually develop under correct conditions. One of the main ANOVA project findings was the fantastic floral response to light treatments consisting of high PPFD spread at the 660 red nanometer specifically.

Here after considering all the different ANOVA factors, and their significance weight we realize that this is a complicated science Botany, and therefore will require multiple phases for more practical results. Clearly there is a place for white light LEDs in the global market, but the real question we are asking is does the white light engineering design optimize spectrum and electrical consumption which will help agriculturists product more affordable food? Below are several spectral measurements that will help one understand light and the different capabilities of each spectrum.

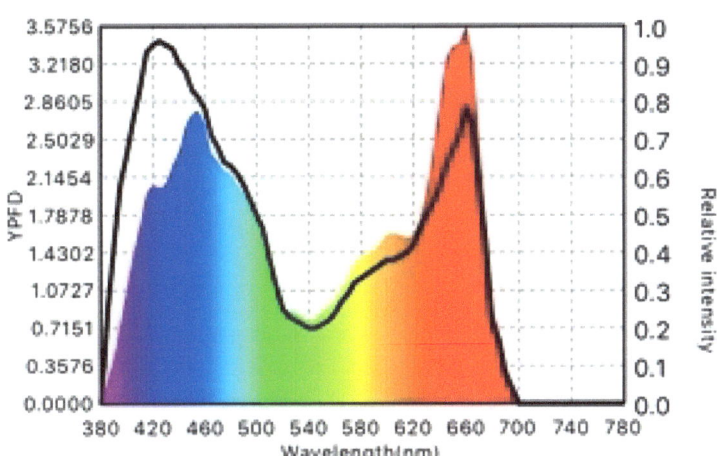

Here we can observe the spectrum of the sum. This measurement was taken On a sunny day in Florida at 3:00PM in the afternoon. Again its worth emphasizing the efficiency of the sun light and the fact that it is a free natural resource. Business models and performance strategies utilizing this aspect are more likely to get funding from venture capitalists.

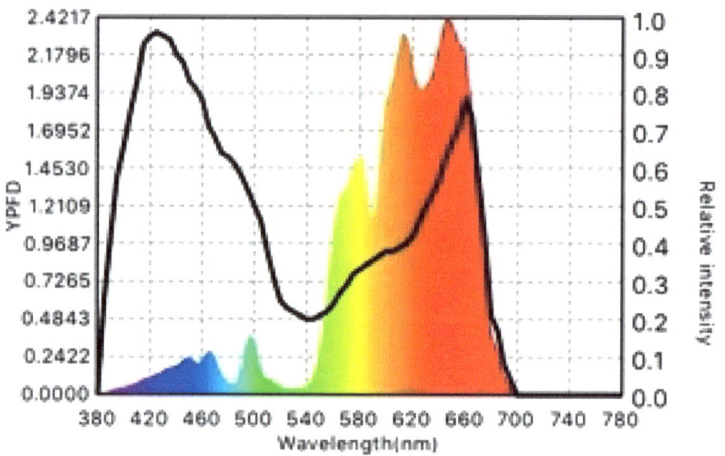

Above we can see a spectrum from a 1000 watt High Pressure Sodium light from Genesis. The area of the medical curve from 580nm to700nm is efficiently filled, although the blue area is lacking strong values. High Pressure Sodium bulbs are the most popular bulb for horticulture applications. As more and more medical aspects are discovered this curve and the blue area of the spectrum will grow in significance.

Weighted Spectrum

Ref.: Medicinal Plant

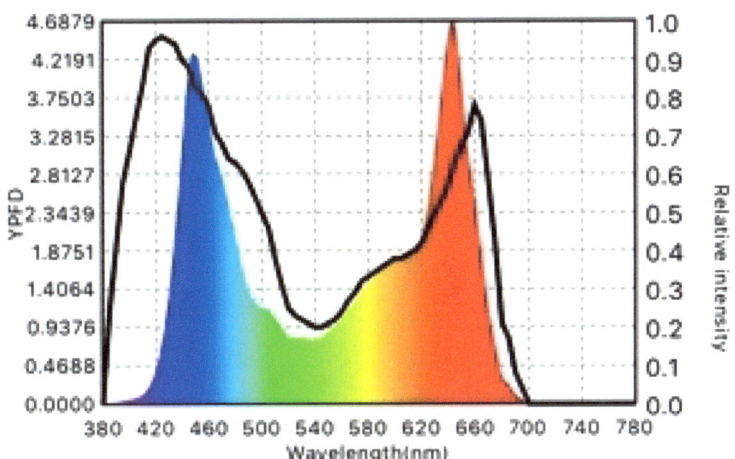

Here is a visual of a white light LED the Solar Eclipse 450 with UVB by AMARE Technology. Here we can see that this is a broader spectrum than the High Pressure Sodium lamp. This white light LED spectrum looks much more like the distribution coming from the sun. We can note that the area of blue covered in the white light LED is much better than the HPS light. Another important note from the previous paper is the inverse square law and how quick the light weakens.

Weighted Spectrum

Ref.: Medicinal Plant

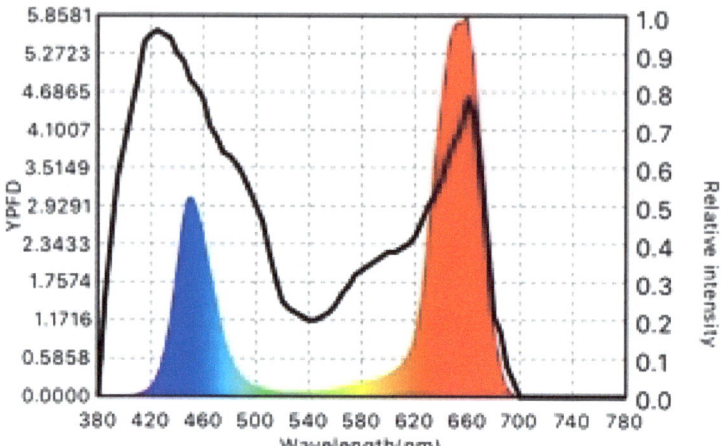

Here is the spectral layout of the Solar System 550 from California Light Works. The blue area is less than the white light LED, however the red area is significantly greater in terms of PPFD and YPFD. And as this research suggests the red 660nm portion of the light spectrum is essential for the onsite of healthy floral formations and fruit development.

Weighted Spectrum

Ref.: Medicinal Plant

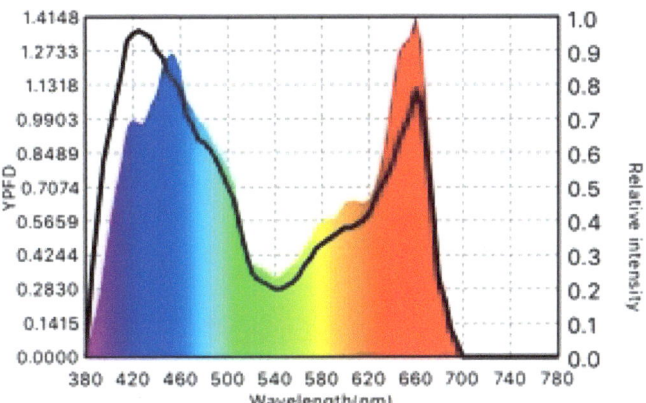

Now let's take another look at the Suns spectrum. The sun is a Blackbody and therefore emits light efficiently and also efficiently absorbs energy. This measurement was also taken at 3:00Pm in the afternoon except for this measurement it was very cloudy outside. We can begin to visualize our atmosphere and it purpose of blocking harmful UVA and some UVB light.

The complete spectrum coming from the Sun including UVB, UVC, Green light 500-540nm and Far red 700-800nm all have effects on plants and play a synergetic role when combined. When we consider indoor horticulture lighting and the fact that resources are used to light these lamps it is not recommended to include the green portion of the spectrum. Although the UVB, UVC and Far Red are recommended to be included in the spectrum.

Throughout the process of the lab several other factors came to mind, and although they are not incorporated into the statistical ANOVA they are worth mentioning here in the final analysis. Are the products marketed honestly? Is eye protection required? The strength of the light at the 660nm specifically? UV end of the spectrum? Finally is the fan quiet?

Another area of controversy is whether or not to adjust the spectrum to optimize vegetative and flowering phases? This research suggests yes this is true.

Generally speaking there is a lot of variety when it comes to LED lighting for horticulture applications. The spectrum is essential, some use a 400-700nm while others use a 380-780nm spectrum. There are companies Like AMARE technology for example that emphasize the importance of UV, and others suggest IR is more important. Clearly there is a lot of research being conducted right now and much more on the horizon. There are LED dot matrixes like the Solar System 550 & G8-900, and others use a different setup for their emitters and drivers like COB lights.

When we consider secondary optics there is also some variation across the industry. Secondary optics is very important as all LEDs are susceptible to the inverse square law, and therefore intensity diminishes with distance. Since this LED ANOVA was designed with the investor in mind, moving toward a PFAL we are very comfortable recommending that the California Light Works Solar System 550 is the most economical fixture out of the ten tested. Please see LED ANOVA, ANOVA R/R, and ANOVA Excel for statistical values.

The Solar System 550 Visual Mode Allows for Easy Plant Observation

Ion 8 Lighthouse Hydro

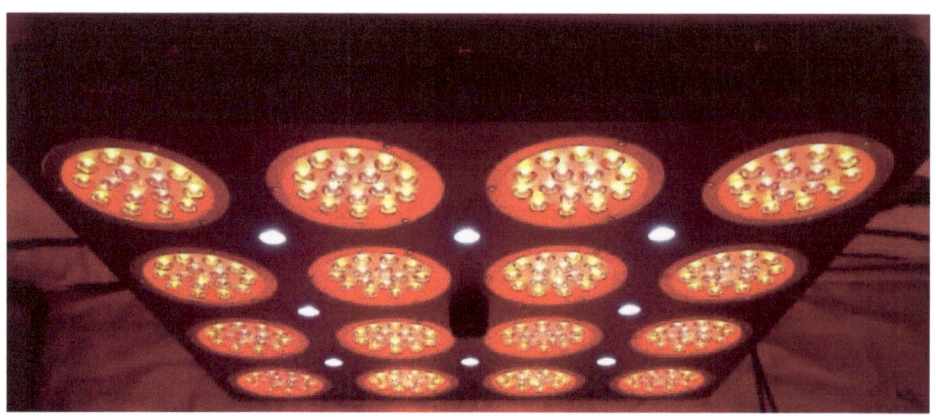

PPFD (400-700)	DLI	YPFD (380-780nm)
856.48	74	781.35
854.01	73.787	777.17
469.58	40.572	443.61
180.13	15.563	170.77
385.37	33.269	364.62
2745.57/5=549	237.218/5=47.443	2537.52/5=507.504

E% Medical Curve	E% McCree Curve	R/B	Temp Rise
47.53	40.28	7	74.7
46.75	40.18	7.33	86.1
28.02	21.82	67.49	86.7
22.26	16.44	68.7	86.2
25.64	19.21	64.69	82.6
170.2/5=34.04	137.93/5-27.586	215.21/5=43.042	86.7-74.7=12

PPFD (400-700nm) average/w draw	549.114/897w
PPFD (400-700nm) average/$	549.114/$2,499
DLI (mol/m^2) average/w draw	47.4436/897w
DLI (mol/m^2) average/$	47.4436/$2499
YPFD (380-780nm) average/w draw	507.504/897w
YPFD (380-780nm) average/$	507.504/$2,499
Average of Efficiency Percentage for McCree Curve	27.59%
Average of efficiency percentage for Medical Curve	34.04%
Average of Red Blue Ratio R/B	43.04%
spectrum adjustment	yes
LED connectivity option	no
Remote control ability via smart phone app	no
Fixture Lifespan in hours	40,000 hrs.
Average Heat Output	12f
Warrantee	3 years
VC funding requirements	no
Weight	42 lbs.
Price/Lifespan	$2499/40,000
Humidity Resistance for greenhouses	no
Rebates from US power companies	no

Solar Pro 900AMARE Technology

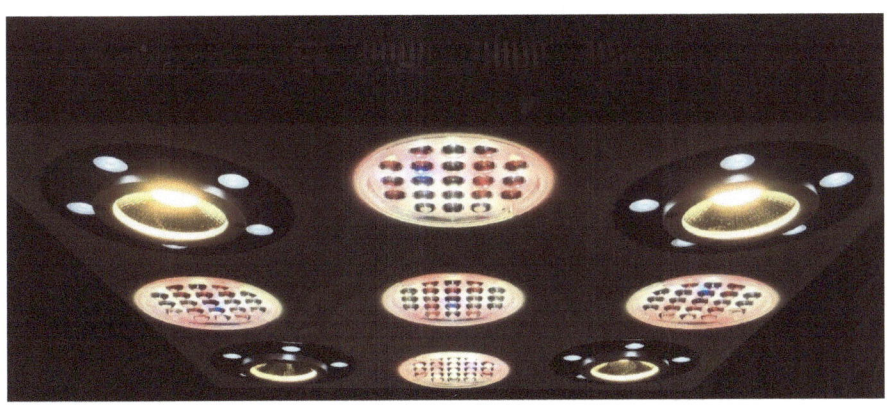

PPFD (400-700)	DLI	YPFD (380-780nm)
1340.2	115.79	1172.9
1346.5	116.33	1177.3
499.18	43.129	448.87
322.63	27.875	290.65
442.01	38.19	397.82
3950.52/5=790.104	341.314/5=68.2628	3487.54/5=697.508

E% Medical Curve	E% McCree Curve	R/B	Temp Rise
71.62	63.18	1.82	79.2
72.24	63.65	1.79	85.8
58.02	55.39	3.45	86.4
56.91	54.71	3.75	86.4
57.86	55.78	3.57	86.2
316.65/5=63.33	292.71/5=58.542	14.38/5=2.876	86.4-79.2=7.2

PPFD (400-700nm) average/w draw	790.104/902w
PPFD (400-700nm) average/$	790.104/$1,995
DLI (mol/m^2) average/w draw	68.2628/902w
DLI (mol/m^2) average/$	68.2628/$1,995
YPFD (380-780nm) average/w draw	697.508/902w
YPFD (380-780nm) average/$	697.508/$1,995
Average of Efficiency Percentage for McCree Curve	58.54%
Average of efficiency percentage for Medical Curve	63.33%
Average of Red Blue Ratio R/B	2.88%
spectrum adjustment	yes
LED connectivity option	no
Remote control ability via smart phone app	no
Fixture Lifespan in hours	50000 hrs.
Average Heat Output	7.2f
Warrantee	5 years
VC funding requirements	yes
Weight	42 lbs.
Price/Lifespan	$1,995/50,000
Humidly Resistance for greenhouses	no
Rebates from US power companies	no

Solar Eclipse 450 uvb AMARE Technology

PPFD (400-700)	DLI	YPFD (380-780nm)
1072.5	92.663	961.69
1052	90.893	922.78
333.59	22.822	296.94
154.46	13.346	138.07
276.13	23.858	245.71
2888.68/5=577.736	243.582/5=48.7164	2565.19/5=513.038

E% Medical Curve	E% McCree Curve	R/B	Temp Rise
63.95	55.27	1.56	72.4
63.84	54.08	1.55	74.3
63.51	65.87	2.33	73.2
61.44	65.52	2.74	72.1
63.74	66.79	2.36	73.9
316.48/5=63.296	307.53/5=61.506	10.54/5=2.108	74.3-72.1=2.2

PPFD (400-700nm) average/w draw	577.736/487w
PPFD (400-700nm) average/$	577.736/$1,108
DLI (mol/m^2) average/w draw	48.7164/487w
DLI (mol/m^2) average/$	48.7164/$1,108
YPFD (380-780nm) average/w draw	513.038/487w
YPFD (380-780nm) average/$	513.038/$1,108
Average of Efficiency Percentage for McCree Curve	61.51%
Average of efficiency percentage for Medical Curve	63.30%
Average of Red Blue Ratio R/B	2.11%
spectrum adjustment	yes
LED connectivity option	no
Remote control ability via smart phone app	no
Fixture Lifespan in hours	50,000 hrs.
Average Heat Output	2.2f
Warrantee	5 years
VC funding requirements	yes
Weight	33 lbs.
Price/Lifespan	$1,108/50,000
Humidity Resistance for greenhouses	no
Rebates from US power companies	no

Universe OGlights-3500K

PPFD (400-700)	DLI	YPFD (380-780nm)
535.8	46.293	479.95
513.98	44.408	459.71
318.67	27.533	285.22
249.58	21.564	223.12
340.25	29.398	304.24
1958.28/5=391.656	169.216/5=33.8432	1752.24/5=350.448

E% Medical Curve	E% McCree Curve	R/B	Temp Rise
59.24	62.36	3.03	73.4
59.89	62.65	2.85	78.3
59.51	62.41	2.91	78.4
59.77	62.54	2.83	78.6
59.87	62.63	2.83	78.8
289.28/5=59.656	312.59/5=62.518	14.45/5=2.89	78.8-73.4=5.4

PPFD (400-700nm) average/w draw	391.656/494w
PPFD (400-700nm) average/$	391.656/$1,099
DLI (mol/m^2) average/w draw	33.8432/494w
DLI (mol/m^2) average/$	33.8432/$1,099
YPFD (380-780nm) average/w draw	350.448/494w
YPFD (380-780nm) average/$	350.448/$1,099
Average of Efficiency Percentage for McCree Curve	62.52%
Average of efficiency percentage for Medical Curve	59.66%
Average of Red Blue Ratio R/B	2.89%
spectrum adjustment	no
LED connectivity option	no
Remote control ability via smart phone app	no
Fixture Lifespan in hours	175,000 hrs.
Average Heat Output	5.4f
Warrantee	no
VC funding requirements	no
Weight	25 lbs.
Price/Lifespan	$1099/175,000
Humidity Resistance for greenhouses	yes
Rebates from US power companies	no

G8-900 DormGrow

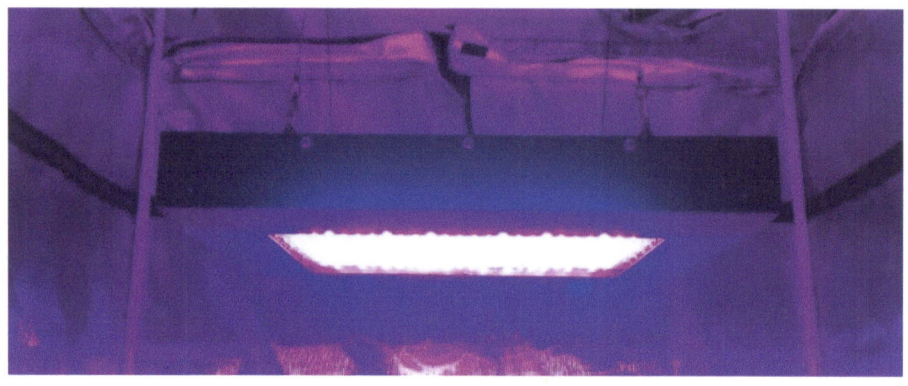

PPFD (400-700)	DLI	YPFD (380-780nm)
699.29	60.419	623.4
702.42	60.689	623.98
279.9	24.183	243.59
154.76	13.371	133.16
199.84	17.266	172.47
2006.21/5=401.242	175.928/5=35.1856	1796.6/5=359.32

E% Medical Curve	E% McCree Curve	R/B	Temp Rise
47.3	35.1	1.89	76.1
48.33	35.98	1.88	80.4
51.89	35.85	1.3	80.6
51.59	33.19	1.06	80.6
51.48	33.76	1.13	80.5
250.59/5=50.118	173.79/5=34.758	7.26/5=1.452	80.6-76.1=4.5

PPFD (400-700nm) average/w draw	401.242/504w
PPFD (400-700nm) average/$	401.242/$1,049
DLI (mol/m^2) average/w draw	35.1856/504w
DLI (mol/m^2) average/$	35.1856/$1,049
YPFD (380-780nm) average/w draw	359.32/504w
YPFD (380-780nm) average/$	359.32/$1,049
Average of Efficiency Percentage for McCree Curve	34.76%
Average of efficiency percentage for Medical Curve	50.12%
Average of Red Blue Ratio R/B	145.20%
spectrum adjustment	no
LED connectivity option	no
Remote control ability via smart phone app	no
Fixture Lifespan in hours	50,000 hrs.
Average Heat Output	4.5f
Warrantee	2 years
VC funding requirements	no
Weight	24.8 lbs.
Price/Lifespan	$1,049/50,000
Humidity Resistance for greenhouses	no
Rebates from US power companies	yes

Solar System 550- California Light Works

PPFD (400-700)	DLI	YPFD (380-780nm)
672.14	58.073	596.31
671.16	57.989	594.3
356.02	30.76	318.88
140.34	12.125	124.25
228.1	19.708	202.76
2067.76/5=413.552	178.755/5=35.751	1836.5/5=367.3

E% Medical Curve	E% McCree Curve	R/B	Temp Rise
39.88	31.31	3.39	76.3
39.66	31.31	3.42	79.7
28.85	20.3	4.76	80.1
32.55	22.51	3.24	79.9
31.01	21.53	3.71	79.7
171.95/5=34.39	126/5=25.392	18.52/5=3.704	80.1-76.3=3.8

PPFD (400-700nm) average/w draw	413.552/405w
PPFD (400-700nm) average/$	413.552/$795
DLI (mol/m^2) average/w draw	35.751/405w
DLI (mol/m^2) average/$	35.751/$795
YPFD (380-780nm) average/w draw	367.4/405w
YPFD (380-780nm) average/$	367.3/$795
Average of Efficiency Percentage for McCree Curve	25.39%
Average of efficiency percentage for Medical Curve	34.39%
Average of Red Blue Ratio R/B	3.70%
spectrum adjustment	yes
LED connectivity option	yes
Remote control ability via smart phone app	no
Fixture Lifespan in hours	50,000 hrs.
Average Heat Output	3.8f
Warrantee	5 years
VC funding requirements	yes
Weight	13 lbs.
Price/Lifespan	$795/50,000
Humidity Resistance for greenhouses	yes
Rebates from US power companies	yes

V1 VIVID GRO

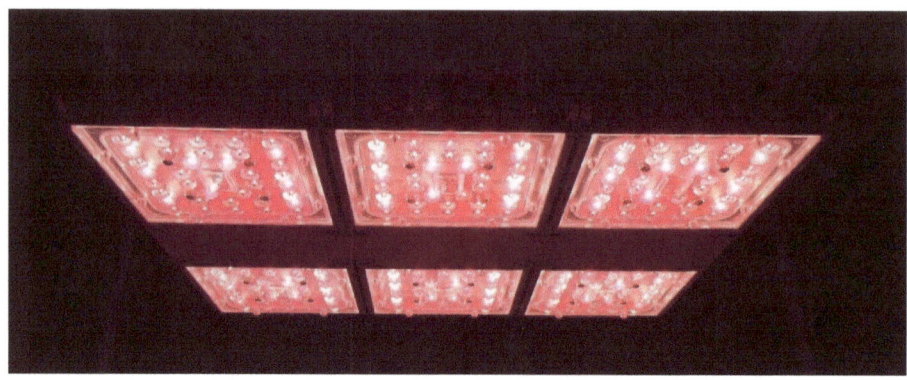

PPFD (400-700)	DLI	YPFD (380-780nm)
300.62	25.973	271.75
298.73	25.81	269.69
257.29	22.229	232.4
254.92	22.025	230.58
177.43	15.33	160.24
1288.99/5=257.798	111.367/5=22.2734	1164.66/5=232.932

E% Medical Curve	E% McCree Curve	R/B	Temp Rise
23.35	15.94	9.05	74.1
23.48	16.02	8.86	75.3
22.2	14.98	9.73	73.6
21.99	14.87	10.67	72.4
22.21	15.03	9.91	72.6
113.23/5=22.646	76.84/5=15.683	48.22/5=9.644	75.3-72.4=2.9

PPFD (400-700nm) average/w draw	257.798/387w
PPFD (400-700nm) average/$	257.798/$1,145
DLI (mol/m^2) average/w draw	22.2734/387w
DLI (mol/m^2) average/$	22.2734/$1,145
YPFD (380-780nm) average/w draw	232.932/387w
YPFD (380-780nm) average/$	232.932/$1,145
Average of Efficiency Percentage for McCree Curve	15.68%
Average of efficiency percentage for Medical Curve	22.65%
Average of Red Blue Ratio R/B	9.64%
spectrum adjustment	no
LED connectivity option	no
Remote control ability via smart phone app	no
Fixture Lifespan in hours	50,000 hrs.
Average Heat Output	2.9f
Warrantee	3 years
VC funding requirements	yes
Weight	40 lbs.
Price/Lifespan	$1,145/50,000
Humidity Resistance for greenhouses	yes
Rebates from US power companies	yes

V2 VIVID GRO

PPFD (400-700)	DLI	YPFD (380-780nm)
626.71	54.147	557.29
629.24	54.367	557.74
377.39	32.606	340.07
294.21	25.42	266.06
461.81	39.901	414.22
2389.36/5=477.872	206.441/5=41.2882	2135.38/5=427.076

E% Medical Curve	E% McCree Curve	R/B	Temp Rise
39.44	29.19	4.65	72.7
38.85	29.42	4.79	78.8
26.95	18.9	8.09	79.2
23.19	15.81	9.33	79.7
30.45	21.95	6.88	79.5
158.88/5=31.776	115.27/5=23.054	33.74/5=6.748	79.7-72.7=7

PPFD (400-700nm) average/w draw	477.552/601w
PPFD (400-700nm) average/$	477.872/$1590
DLI (mol/m^2) average/w draw	41.2882/601w
DLI (mol/m^2) average/$	41.2882/$1590
YPFD (380-780nm) average/w draw	427.076/601w
YPFD (380-780nm) average/$	427.076/$1590
Average of Efficiency Percentage for McCree Curve	23.05%
Average of efficiency percentage for Medical Curve	31.78%
Average of Red Blue Ratio R/B	674.80%
spectrum adjustment	no
LED connectivity option	no
Remote control ability via smart phone app	no
Fixture Lifespan in hours	50000 hrs.
Average Heat Output	7f
Warrantee	5 years
VC funding requirements	yes
Weight	64 lbs.
Price/Lifespan	$1590/50,000
Humidity Resistance for greenhouses	yes
Rebates from US power companies	yes

COB Grow Lights – Scorpion9

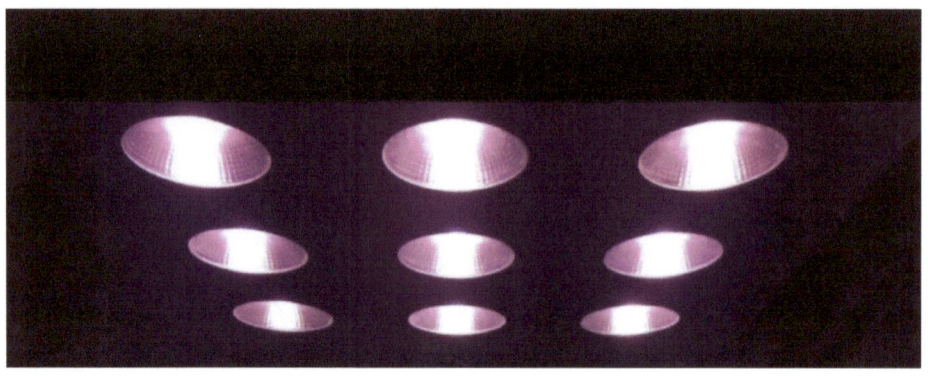

PPFD (400-700)	DLI	YPFD (380-780nm)
857.32	74.072	781.37
849.72	73.416	774.67
218.84	18.908	202.16
104.03	8.9883	96.221
165.15	14.269	152.66
2,195.06/5=439.012	189.6533/5=37.93066	2007.081/5=401.4162

E% Medical Curve	E% McCree Curve	R/B	Temp Rise
56.61	53.31	2.46	73.1
56.22	52.86	2.45	74.5
47.08	44.06	2.77	73.8
46.72	43.86	2.87	73.7
46.95	44.06	2.82	74.1
253.58/5=50.716	238.15/5=47.63	13.37/5=2.674	74.5-73.1=1.4

PPFD (400-700nm) average/w draw	439.012/610w
PPFD (400-700nm) average/$	439.012/$799
DLI (mol/m^2) average/w draw	37.93066/610w
DLI (mol/m^2) average/$	37.93066/$799
YPFD (380-780nm) average/w draw	401.4162/610w
YPFD (380-780nm) average/$	401.4162/$799
Average of Efficiency Percentage for McCree Curve	47.63%
Average of efficiency percentage for Medical Curve	50.72%
Average of Red Blue Ratio R/B	2.67%
spectrum adjustment	no
LED connectivity option	no
Remote control ability via smart phone app	no
Fixture Lifespan in hours	50,000 hrs.
Average Heat Output	1.4f
Warrantee	no
VC funding requirements	yes
Weight	38 lbs.
Price/Lifespan	$799/50,000
Humidity Resistance for greenhouses	yes
Rebates from US power companies	no

Blackbody

When we consider the sun as a blackbody there are tow things we must understand; first as the temperature of a blackbody increase peak emission moves tward shorted wavelengths. This will be explained further in Wien's Displacement Law Since light has two major aspects and consistently moves at 186,000 miles per second aka the speed of light, its recommended that we frequently regress to observing the sun, a blackbody that all LED lights are trying to mimic. Both aspects of light move together at this speed although the two different aspects which is wave aspect aka electromagnetic waves and particle aspect or proton both move in a perpendicular fashion at 186,000 miles per second. The wave aspect and particle aspect have a continuous inverse relationship. Blackbody- "A blackbody (sometimes spelled "black body") is a theoretically ideal radiator and absorber of energy at all electromagnetic wavelength s. The term comes from the fact that a cold blackbody appears visually black. The energy emitted by a blackbody is called blackbody radiation. This takes the form of an electromagnetic field having an intensity-versus-wavelength relation whose graph looks like a skewed, bell-shaped statistical curve. The maximum point on the curve shows the wavelength at which the radiation intensity is greatest. This wavelength depends on the thermodynamic temperature , in kelvin s, of the object. The higher the temperature, the shorter the wavelength at which the radiation is most intense. The wavelength and temperature are related by a function involving [Wien's constant](). Scientists attempt to determine the temperatures of distant objects in space by observing their blackbody radiation. The calculations are made by assuming that celestial objects behave as perfect blackbodies. A blackbody is a theoretical ideal, but many astronomical objects come reasonably close to this ideal."

http://whatis.techtarget.com/definition/blackbody

Florida Sunset

Star Temperatures

Stars approximate blackbody radiators and their visible color depends upon the temperature of the radiator. The curves show blue, white, and red stars. The white star is adjusted to 5270K so that the peak of its blackbody curve is at the peak wavelength of the sun, 550 nm. From the wavelength at the peak, the temperature can be deduced from the Weins Displacement Law

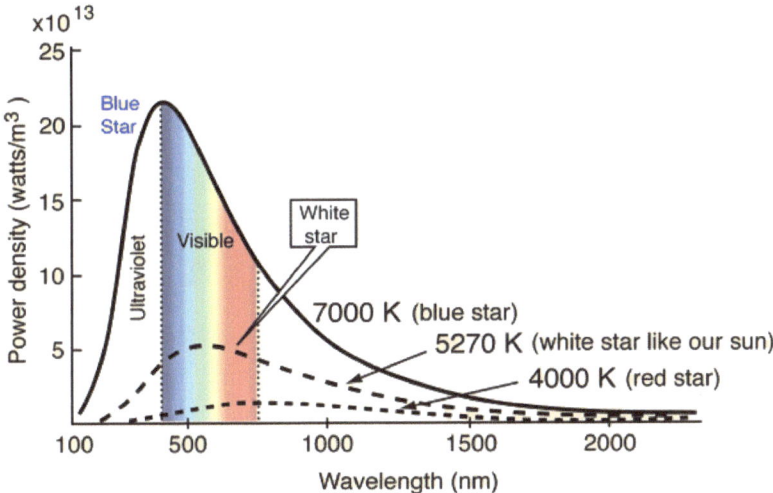

Wien's Displacement Law

"When the temperature of a blackbody radiator increases, the overall radiated energy increases and the peak of the radiation curve moves to shorter wavelengths. When the maximum is evaluated from the Planck radiation formula, the product of the peak wavelength and the temperature is found to be a constant."

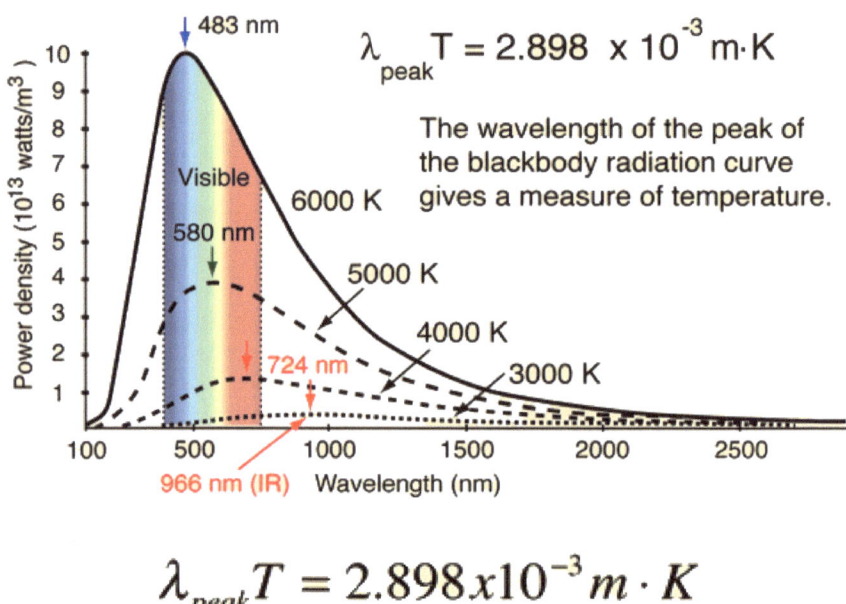

$\lambda_{peak} T = 2.898 \times 10^{-3} \, m \cdot K$

The wavelength of the peak of the blackbody radiation curve gives a measure of temperature.

$$\lambda_{peak} T = 2.898 \times 10^{-3} m \cdot K$$

"This relationship is called Wien's displacement law and is useful for the determining the temperatures of hot radiant objects such as stars, and indeed for a determination of the temperature of any radiant object whose temperature is far above that of its surroundings."

"It should be noted that the peak of the radiation curve in the Wien relationship is the peak only because the intensity is plotted as a function of wavelength. If frequency or some other variable is used on the horizontal axis, the peak will be at a different wavelength."

V2 Lighting Science

Radiation Curves

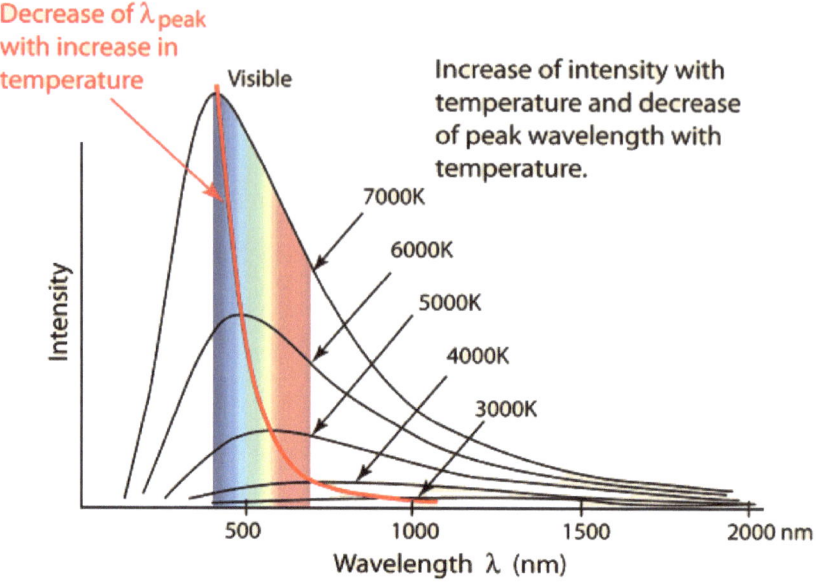

"The wavelength of the peak of the blackbody radiation curve decreases in a linear fashion as the temperature is increased (Wien's displacement law). This linear variation is not evident in this kind of plot since the intensity increases with the fourth power of the temperature. The nature of the peak wavelength change is made more evident by plotting the fourth rout of the intensity."

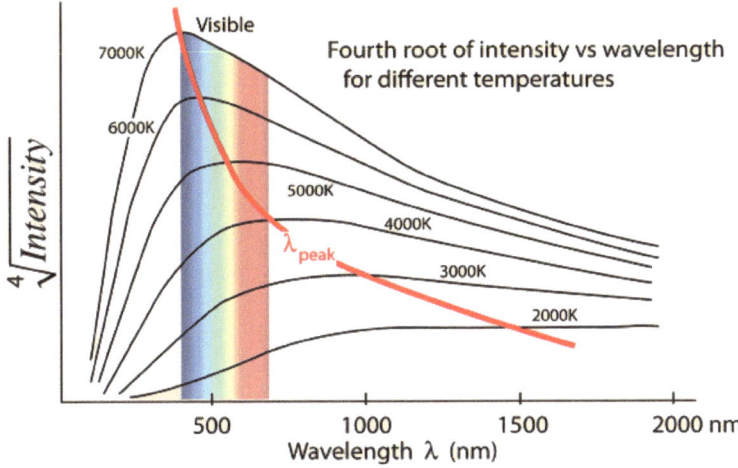

"The wavelength of the peak of the blackbody radiation curve decreases with increasing temperature according to Wien's displacement law. The fourth root of the intensity shows the variation of the wavelength more clearly than the plot of the full radiated intensity."

http://hyperphysics.phy-astr.gsu.edu/hbase/wien.html

UV Radiation

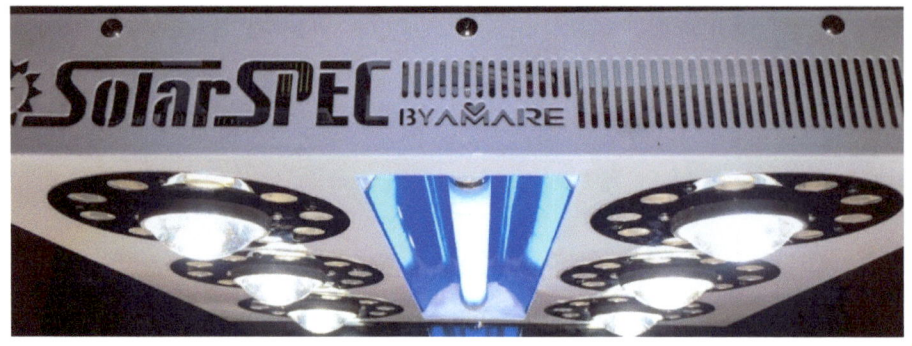

UVA 320-400nm/ UVB 290-320nm/ UVC 200-290nm

Most UV light is absorbed by our ozone layer, most specifically all UVC, most UVB, and some UVA. This UV light is the reason we wear sun screen when we go to the beach or out on the water. This portion of the spectrum can cause cataracts in the eye and skin cancer like Basil cell carcinoma- "Basal cell carcinoma is a cancer that grows on parts of your skin that get a lot of sun. It's natural to feel worried when your doctor tells you that you have it, but keep in mind that it's the least risky type of skin cancer. As long as you catch it early, you can be cured."
http://www.webmd.com/melanoma-skin-cancer/basal-cell-carcinoma#1

https://www.sparkfun.com/products/8662

http://www.uvled.us/uv-led

http://www.mouser.com/ultraviolet_leds/

http://www.lumileds.com/products/uv-leds

Ultraviolet (UV) is an electromagnetic radiation with a wavelength from 10 nm (30 PHz) to 400 nm (750 THz), shorter than that of visible light but longer than X-rays. UV radiation constitutes about 10% of the total light output of the Sun, and is thus present in sunlight. It is also produced by electric arcs and specialized lights, such as mercury-vapor lamps, tanning lamps, and black lights. Although it is not considered an ionizing radiation because its photons lack the energy to ionize atoms, long-wavelength ultraviolet radiation can cause chemical reactions and causes many substances to glow or fluoresce. Consequently, the biological effects of UV are greater than simple heating effects, and many practical applications of UV radiation derive from its interactions with organic molecules.

https://en.wikipedia.org/wiki/Ultraviolet

Indeterminate Trellising

Additional thoughts from LED industry professionals.

"Pure white lights bleed into yellow and green as a result of the phosphor coating over the LEDs. So I wouldn't say it is intentional in most cases. One could argue that you are wasting energy with that kind of spectrum much like many LED manufacturers claim HPS lights waste energy on the large amounts of orange and yellow they produce. Blue and red LEDs are the most cost effective to produce. White LEDs are actually just phosphor coated blue LEDs, so naturally they are slightly more expensive than blue LEDs."

"As far as the importance of the 500-560nm spectrum. It isn't necessary. It's been proven that you can get great growth rate and results with only 440-450 paired with ~660. This is most relevant with high turnover low dollar crops like leafy greens, and micro greens."

"With light intensive crops like Cannabis, I believe there is a benefit to include the 500-560nm spectrum. Though the benefits are still unclear, and debated. Some say it helps increase the penetration of the other spectra."

Dylan Cirrus

"White LEDs are basically phosphor covered blue LEDs, so you have losses compared to the blue LED but still a better efficiency than discrete colors in-between red and blue. Use red and blue in for example a greenhouse as supplemental, white ones for a more wide spectrum indoor, combined possibly with red, far red and blue.

"There is white and there is white. You won't see the difference, a plant will. Your lcd screen shows only red, green and blue yet you perceive it as white together. A plant doesn't: it sees three distinct peeks. Same with phosphor spectra."

Theo Gatvia

"White light created by phosphor coated blue LEDs has approximately 1/3 the PAR output of the white light created by our RGB LEDs. It's for this reason the production is incredibly weak."

Cammie HydrogrowLED

"From our point of view, the main advantage of LEDs for horticulture is the ability to vary the spectrum to match the season, the type of plant, the morning to night schedule, etc. to best match natural sunlight and maximize growth and yield. Fixed spectrum LEDs eliminate this key feature. It is much simpler and cheaper to just use fixed spectrum LEDs which is at best a compromise solution. This is similar to the solution used with fixed spectrum florescent lighting."

Craig CLW

Also we should all take note that the leading LED and lighting companies in general across the world are focusing Research and Development into LED light bars that can be adjusted and connected as the grower desires. Also a feasible application is using these light bars in-between rows in a greenhouse or a indoor facility.

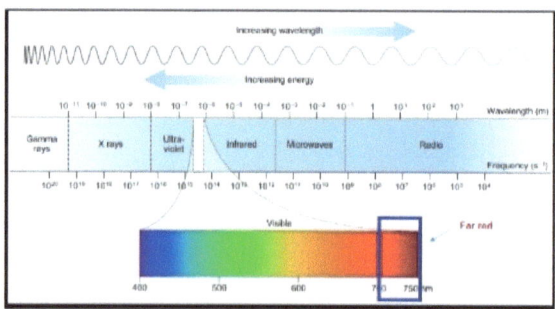

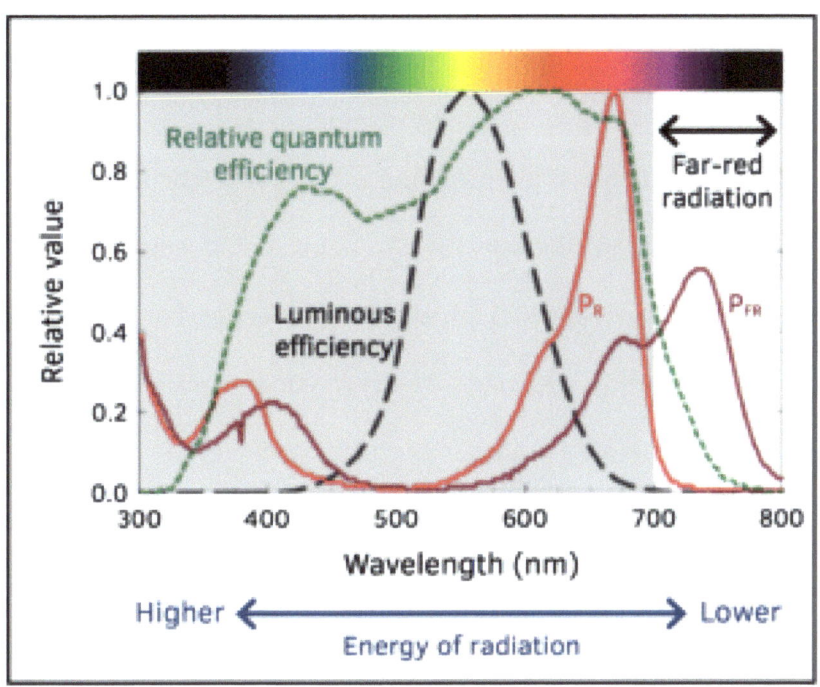

Figure 1. Far-red radiation is barely visible to us (as shown by the luminous efficiency curve) but can have big impacts on flowering and extension growth of plants.

V1 Lighting Science

Far Red Radiation

The far red area of the spectrum which is 700-800nm is also important to plants. This is possible why High Pressure Sodium bulbs still are successful flowering fixtures. Far red is not absorbed by leaves, but instead is transmitted through the leaves. Red radiation gets absorbed by chlorophyll a and b, for example a branch in the shade or under a canopy still received far red (not red) radiation due to transmission and therefore is able to recognize it's shaded from the full spectrum. This is the shade avoidance response through Phfr and perhaps including a far red radiation option will be possible moving forward?

Far red radiation 700-800nm is essential for the plants morphology specifically growth including floral structures, size of leaves, intermodal distance and shape of different plant structures. Far red radiation is necessary for the proper <u>morphology of plants-</u> *"Plant morphology or phytomorphology is the study of the physical form and external structure of plants. This is usually considered distinct from plant anatomy, which is the study of the internal structure of plants, especially at the microscopic level."*

<u>https://en.wikipedia.org/wiki/Plant_morphology</u>

<u>Floral Morphology</u>

"The above figure shows the far red waveband relative to the spectrum of light that controls growth and development of plants 300-800nm. Far red promotes expansion growth meaning it influences the size of leaves, the length of stems, and ultimately the height of plants. One of the ways plants perceive light quality is through the pigment phytochrome, which exists in two different forms. One form is called red-absorbing form Pr and the other is the far red absorbing Pfr. Sunlight emits almost as much far red radiation as red light but reflect or transmit most far red." This again is the shade avoidance response, a critical mechanism of a plants metabolism, photosynthesis, growth rate, and therefore biomass."

http://www.gpnmag.com/article/a-closer-look-at-far-red-radiation/

http://www.gpnmag.com/article/r-fr-ratio/

http://www.rapidled.com/cree-xp-e-far-red-led/

http://leds.hrt.msu.edu/assets/Uploads/Univ.-of-Arizona-Greensys-FR-EOD-presentation-2011.pdf

http://www.lumigrow.com/products/lumibulb-fr-far-red-led-grow-bulb/?_kk=red%20LED%20grow%20bulb&_kt=0d04acd4-1089-43a6-ac9f-a7a604b56f63

https://www.monstergardens.com/powerPAR-LED-Bulb---Far-Red-15W-E27

http://www.ledengin.com/files/products/LZ1/LZ1-00R302.pdf

http://growlightsource.com/how-to-accelerate-bud-production-and-quality-with-the-far-red-730n-flower-initiator/

"The relative quantum efficiency curve in Figure 1 represents the effect of radiation at promoting photosynthesis on an instantaneous basis. Adding far red to a light spectrum can increase leaf size enabling plants to capture more light and potentially increase growth. Therefore, over time, far red radiation can indirectly increase growth." We feel that including a far red radiation aspect into a LED grow light is the best recommendation and will further increase the competitive capabilities of LED companies relative to HPS bulbs.

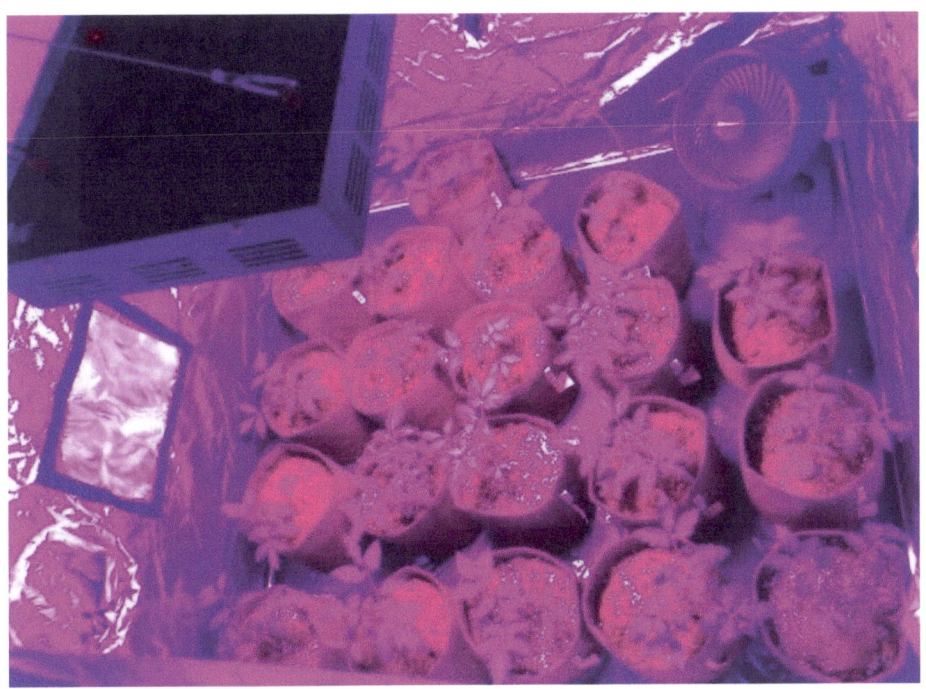

The Far Red spectrum of the G8-900

Clearly each LED manufacturer has their own methods of designing a spectrum, and how they choose to manufacture their fixtures. Out of the Red/Blue LED lights tested in this research project the G8-900 was the only Red/Blue fixture to incorporate a **735nm** far red emitters into their light.

"The combination of 660 nm and 730-735 nm deep red light creates the Emerson effect and results in an even higher rate of photosynthesis in plants."

http://activegrowled.com/about/

"Light intensity has a much larger effect on plant growth than light spectrum, and this is especially true when electrical lighting supplements sunlight, such as in greenhouse applications."

https://news.gov.bc.ca/stories/century-of-farming-research-celebrated-in-agassiz

http://www.red-raspberry.org/presentations/Canada2012RO_BCPerspective.pdf

http://www.agr.gc.ca/eng/science-and-innovation/research-centres/british-columbia/agassiz-research-and-development-centre/scientific-staff-and-expertise/forge-thomas-phd/?id=1181937326469

Light is a source of energy and information for plants. It's needed as energy in photosynthesis and it provides plants critical information about its environment, which the plant needs in order to germinate, grow to a certain size or shape, induce protective substances, flower and when to change from vegatative growth. Plants react to quality, intensity, duration and the direction of light.

"In addition to the light visible to humans (400 nm – 700 nm) plants "see" or use other radiation too. The 400 nm–700 nm wavelength range is called "Photosynthetically Active Radiation" or PAR. Much of the light that plants need is in this range, but for optimal growth result, UV light (280 - 400 nm) and/or far-red light (700-800 nm) might be important. For example far-red is critical for the flowering of many plants. All light is not equal to plants, ie. some areas are more important than others."

"Known photoreceptors are most efficient in the blue and red area of the spectrum. Green plants reflect a significant part of light in the green area of the light spectrum, while absorbing a higher percentage of blue and red light. The graph below shows the light spectrum of the sun on a cloudless day."

Veg Development under the Solar System 550

"Much research has been conducted regarding the optimal light spectrum of plants. A good description of this is the Relative Quantum Efficiency curve for plants. It considers the photosynthetic rate of the plant (by measuring CO_2 uptake), the energy of light at different wavelengths and the plants absorption of light ie. what stays in the leaf and is not reflected away or transmitted through it. However, producing a plant which is sellable is not only about photosynthesis. Also other aspects like shape, flowering, color of leaves, color of flowers, taste, smell, root development, etc. are important to have a high quality plant."

Spectral Science Lab

Producing high quality plants with LED lighting

<u>To produce a high quality plant, it is not enough to use just red and blue light typically found in conventional horticultural LEDs. The same applies for High Pressure Sodium lights, which are dominantly yellow.</u>

"A good balanced ratio between red and blue light, needs to be complemented with far-red and green in right proportions for the right applications. For example far-red, 700–800 nm is critical for flowering of many plants. Less is known about the green, 500–600 nm area, but recent research is finally able to conclude that this is also important for plant development. Besides being potentially harmful, plants also show photomorphogenic responses to UV-B (280-315 nm) radiation. Both UV-B and UV-A (315-400 nm) radiation are important in inducing production of phenolics, anthocyanins (coloration) and antioxidants in plants. Thus the full spectrum from about 350–750 nm at a suitable light intensity is interesting in plant cultivation. The key is then to create a light which is optimally suitable for its task, by balancing the different areas so that the plant gets the right energy and signals to achieve the growers' goals."

http://www.valoya.com/plants-light

http://leds.hrt.msu.edu/assets/Uploads/PowerPoints/LEDs-during-vegetable-propagation.pdf

http://leds.hrt.msu.edu/assets/Uploads/PowerPoints/2012-OFA-Short-Course-presentation-Mitchell-Lopez-Runkle-and-Burr.pdf

European Leaders

https://www.heliospectra.com/

http://www.lemnis-oreon.com/nl/home.html

http://www.valoya.com/

Ripe F1 Trust Tomato

*"I want to say that I think you have done a terrific job with your research project and I
am happy that we participated."*

Dorm *Grow LED*

"Given your background and understanding of the science behind grow lights and actual testing experience of leading LED fixtures, including ours, it's a no-brainer on the benefits of you joining our team."

Victor AMARE Technologies

"We would love to work with you on additional research and testing projects.

Let's discuss how this might be organized."

Craig California Light Works

Day Star Life Center

"All of the produce harvested from the ANOVA project will be donated to the Daystar community Life center here in St. Petersburg. A understanding of the surrounding environment along with a helping hand in the community will be a foundational principle of the organization which will allow students to see life off of campus and help them establish a more realistic view of the community and the people living here in St. Pete."

Thanks so much Pennperlite for your generous contribution to this research project. Penn perlite donated all the perlite needed for this first phase ANOVA trial, and we are very pleased to continue working and colaborating with this company due to its commitment to sustainability and clean energy, a great partner for those who share a passion for these foundational principles.

"Pennsylvania Perlite Corporation manufacturers Perlite for the Industrial, Construction, Horticultural, Coatings, Pharmaco-Chemical, Food, and other industries. They offer a complete service for the cryogenic industry with installation, removal, disposal and recycle of Perlite."

http://pennperlite.com/index.html

Root pouches made this project run smoothly and effectively. These products drain much easier that traditional plastic pots, and are made from recycled grocery bags which is a very innovative way of recycling. "Root Pouch excelled in all categories from health of roots, weight of leaf, size of stem branching and trunk size, with least amount of root circling and spiraling. With superior temperature protection from heat and cold."

http://rootpouch.com/nurseries

Foss Hill Road Albion, Maine U.S.A. 04910

Johnny's seeds was kind enough to make a donation of all the F1 Trust tomato seeds that we needed for this ANOVA project, so just wanted to say thank you Johnney's seeds for all your hard work for the gardening community.

http://www.johnnyseeds.com/

All of the additional supplementary products required were supplied by our friends at Microblife. They have done a great job in supplying some of the best and most affordable supplementary products for the market including foliar spray and root dip, a great vitamin blend, and photosynthesis plus.

"Ecological Laboratories, Inc. products and technologies are backed and supported by a modern well equipped highly sophisticated R&D laboratory. It is equipped with the latest evaluation methodologies, and staffed by PhD and Masters Level Microbiologists, Chemists, and highly qualified laboratory assistance. Supporting lab equipment such as microscopes, chemical determination units, and necessary laboratory protective hoods. Operating in a protected laboratory environment with positive air pressure, and HEPA filter recirculated air, offering a protected environment for work and R&D projects and evaluations."

https://www.microbelifehydro.com/

Work Cited

www.ibm.com/marketplace/cloud/statisticalanalysisandreport

http://www-03.ibm.com/software/products/en/spss-advanced-stats

"Statistics for Business and Economics Anderson/Sweeney/Williams"

http://www.valoya.com/products

http://www.genesisbulbs.com/research/

https://fluence.science/science/how-to-compare-grow-lights/

https://fluence.science/science/par-ppf-ppfd-dli/

http://www.leoledgrow.eu/en/measure-par-ppfd

http://smartgrowtechnologies.com/plant-light/

Hopkins, William G. and Norman P. A. Hüner. (2009) Introduction to Plant Physiology. 4nd ed. New York, NY: John Wiley & Sons.

Knight, Rebecca. "Modeling the Unknown – Why Light Calculations are So Hard" https://www.illumitex.com/modeling-unknown-light-calculations-hard-779.

Kozai, T. (2013). "Resource use efficiency of closed plant production system with artificial light: Concept, estimation and application to plant factory." Proceedings of the Japan Academy, Series B 89(10): 447-461.

www.ingramcontent.com/pod-product-compliance
Lightning Source LLC
Chambersburg PA
CBHW041100180526
45172CB00001B/48